Looking at Clouds

Wispy Cirrus Clouds

By Kim Thompson

TABLE OF CONTENTS

A Crabtree Seedlings Book

Crabtree Publishing
crabtreebooks.com

School-to-Home Support for Caregivers and Teachers

This book helps children grow by letting them practice reading. Here are a few guiding questions to help the reader build comprehension skills. Possible answers appear here in red.

Before Reading:

- What do I know about clouds?
  - *I know that the sky has different clouds at different times.*
  - *I know that some clouds are storm clouds.*
- What do I want to learn about clouds?
  - *I want to learn about different kinds of clouds.*
  - *I want to know what the weather will be like by looking at clouds.*

During Reading:

- What have I learned so far?
  - *I have learned that clouds are made of water droplets or ice crystals.*
  - *I have learned that cirrus clouds are white and wispy.*
- I wonder why...
  - *I wonder why clouds form at different heights.*
  - *I wonder what happens when an airplane flies through clouds.*

After Reading:

- What did I learn about cirrus clouds?
  - *I have learned that cirrus clouds are the highest clouds in the sky.*
  - *I have learned that cirrus clouds can mean that the weather will change in about a day.*
- Read the book again and look for the glossary words.
  - *I see the word **wispy** on page 4 and the word **crystals** on page 8. The other vocabulary words are found on page 22.*

Wispy Cirrus Clouds

Clouds are made of tiny droplets of water or ice. One kind of cloud is called cirrus.

Cirrus clouds are white and **wispy**. They look like thin, flowing waves or curls.

Cloud Watch

Cirrus clouds are sometimes called mares' tails. A mare is a female horse.

Water heated by the Sun turns into an invisible gas called **water vapor**. The gas rises up high. It cools and turns back into water droplets. They join together to make clouds.

Cloud Watch

Water moves from oceans, up to the sky to form clouds, then back down to the ground and the oceans. This is called the water cycle.

The water vapor rises into the sky and cools down. The gas turns back into tiny droplets of water. This change is called condensation.

Cirrus clouds are the highest clouds in the sky. It is very cold way up in the air. Droplets of water freeze. That is why cirrus clouds are made of ice **crystals**.

Cloud Watch

Cirrus clouds form over six miles (10 kilometers) above the ground. That is higher than most airplanes fly!

Cirrus clouds have feathery edges. That is because strong winds high in the sky spread out the ice crystals at the cloud's edges.

Cloud Watch

You can tell what direction the wind is blowing by the shape of a cirrus cloud.

A cloud's droplets fall when they get too heavy to float. This is called **precipitation**. Rain and snow are kinds of precipitation.

All clouds move and change. Warm air can turn the ice crystals in a cirrus cloud back into a gas. The cloud can disappear before your eyes!

Cirrus clouds mean nice, sunny weather. But watch out! They can also be a sign that the weather will change in about a day.

Cloud Watch

A large number of cirrus clouds may appear before stormy weather.

There are different kinds of cirrus clouds. Cirrostratus clouds cover the sky like a thin, white sheet. They can look like a **halo** around the Sun or the Moon.

Cloud Watch

If you see cirrostratus clouds, a rainstorm or snowstorm may come in the next day or two.

Cirrocumulus clouds appear in wavy patches. They may look like scales on a fish. A sky filled with cirrocumulus clouds is sometimes called a **mackerel** sky.

Cloud Watch

Cirrocumulus clouds often form on cold, sunny winter days. They may be a sign that good weather will continue.

Glossary

crystals (KRIS-tls): Small pieces of ice that float in the air in cold weather

halo (HAY-loh): A circle or arc of light surrounding an object

mackerel (MAK-er-uhl): A fish often caught for food

precipitation (pri-sip-i-TAY-shuhn): The falling of water from the sky in the form of rain, sleet, hail, or snow

water vapor (WAW-tur VAY-pur): Water that has changed from a liquid into a gas

wispy (WIS-pee): In long and thin pieces; streaky

Pronunciation Guide

cirrocumulus (sir-oh-KYOO-myuh-luhs)
cirrostratus (sir-oh-STRAY-tuhs)
cirrus (SIR-uhs)

Cloud levels		Cloud types
High clouds *Cloud names that begin with "cirro" are the highest clouds in the sky.*	*above 18,000 feet (5,486 meters)*	**High clouds** *cirrus* *cirrostratus* *cirrocumulus*
Middle clouds *Clouds in the middle of the sky have names that start with "alto."*	*6,500 to 18,000 feet (1,981 to 5,486 meters)*	**Middle clouds** *altostratus* *altocumulus*
Low clouds *Clouds that form low in the sky have "strato" in their name.*	*up to 6,500 feet (1,981 meters)*	**Low clouds** *stratus* *stratocumulus* *nimbostratus*

Index

Written by: Kim Thompson
Designed by: Samara Parent
Series Development: James Earley
Proofreader: Kathy Middleton
Educational Consultant: Marie Lemke M.Ed.
Photographs: Abinieks/Shutterstock.com: p.10;
All other images from Shutterstock
Illustrations: Kathy Kantor: p.7

Crabtree Publishing

crabtreebooks.com 800-387-7650

Printed in the USA/CP2025

Published in Canada
Crabtree Publishing
616 Welland Avenue
St. Catharines, Ontario
L2M 5V6

Published in the United States
Crabtree Publishing
347 Fifth Avenue
Suite 1402-145
New York, New York, 10016

Library and Archives Canada Cataloguing in Publication
Available at Library and Archives Canada

Library of Congress Cataloging-in-Publication Data
Available at the Library of Congress

Hardcover: 978-1-0398-8122-8
Paperback: 978-1-0398-8483-0
Ebook (pdf): 978-1-0398-8242-3
Epub: 978-1-0398-8362-8
Read-Along: 978-1-0398-8567-7
Audio: 978-1-0398-8525-7